# BEHOLD THIS COMPOST

How City-Wide Compost Programs Work and Why We Need Them Now, More Than Ever.

Alex Ulysses Nickel

*Dedicated to the Mountain School Class of Fall 2018*

**TABLE OF CONTENTS**

*Part 1:* An Introduction

*Part 2:* Why We Need City-Wide Compost Programs

*Part 3:* How City-Wide Compost Programs Work

*Part 4:* What Makes San Francisco's Compost Program So Great?

*Part 5:* An Appendix

# This Compost
## By Walt Whitman

### 1

Something startles me where I thought I was safest,
I withdraw from the still woods I loved,
I will not go now on the pastures to walk,
I will not strip the clothes from my body to meet my lover the sea,
I will not touch my flesh to the earth as to other flesh to renew me.

O how can it be that the ground itself does not sicken?
How can you be alive you growths of spring?
How can you furnish health you blood of herbs, roots, orchards, grain?
Are they not continually putting distemper'd corpses within you?
Is not every continent work'd over and over with sour dead?

Where have you disposed of their carcasses?
Those drunkards and gluttons of so many generations?
Where have you drawn off all the foul liquid and meat?
I do not see any of it upon you to-day, or perhaps I am deceiv'd,
I will run a furrow with my plough, I will press my spade
through the sod and turn it up underneath,
I am sure I shall expose some of the foul meat.

### 2

Behold this compost! behold it well!
Perhaps every mite has once form'd part of a sick person—yet behold!
The grass of spring covers the prairies,
The bean bursts noiselessly through the mould in the garden,
The delicate spear of the onion pierces upward,

*The apple-buds cluster together on the apple-branches,*
*The resurrection of the wheat appears with pale visage out of its graves,*

*The tinge awakes over the willow-tree and the mulberry-tree,*
*The he-birds carol mornings and evenings while the she-birds sit on their nests,*
*The young of poultry break through the hatch'd eggs,*
*The new-born of animals appear, the calf is dropt*
*from the cow, the colt from the mare,*
*Out of its little hill faithfully rise the potato's dark green leaves,*
*Out of its hill rises the yellow maize-stalk, the lilacs bloom in the dooryards,*
*The summer growth is innocent and disdainful above all those strata of sour dead.*

*What chemistry!*
*That the winds are really not infectious,*
*That this is no cheat, this transparent green-wash of*
*the sea which is so amorous after me,*
*That it is safe to allow it to lick my naked body all over with its tongues,*
*That it will not endanger me with the fevers that have deposited themselves in it,*
*That all is clean forever and forever,*
*That the cool drink from the well tastes so good,*
*That blackberries are so flavorous and juicy,*
*That the fruits of the apple-orchard and the orange-*
*orchard, that melons, grapes, peaches, plums, will*
*none of them poison me,*
*That when I recline on the grass I do not catch any disease,*
*Though probably every spear of grass rises out of what was once a catching disease.*

*Now I am terrified at the Earth, it is that calm and patient,*
*It grows such sweet things out of such corruptions,*
*It turns harmless and stainless on its axis, with such*
*endless successions of diseas'd corpses,*
*It distills such exquisite winds out of such infused fetor,*
*It renews with such unwitting looks its prodigal, annual, sumptuous crops,*
*It gives such divine materials to men, and accepts such leavings from them at last.*

# Part 1: An Introduction

*Or, how I came to care about compost enough to write a book on it.*

"What's the most beautiful thing you've ever seen?"

This question has been on my mind ever since a biology teacher at my school asked me it while we were rafting down a river north-east of Sacramento with seven of my classmates.

**Fig. 1.** *Me, at the front of the raft, having the time of my life.*

How exactly did I get myself into that situation?

At the beginning of every school year, my entire grade goes on a trip in the hopes that we bond as a class. In eleventh grade, my class and I went camping in Big Basin; in tenth grade, we went kayaking on the Russian River; and, in ninth grade, we went to Donner Pass. Donner Pass, in my humble opinion, is a rather odd place to start my high school experience, seeing as Donner Pass is famous for being the spot the Donner Party — a group of people who traveled along the Oregon Trail and migrated west in the 1840s — decided to resort to cannibalism because they were surrounded by snow and had run out of food. Nonetheless, it was quite an enjoyable trip, and (luckily) none of us had to follow in the footsteps of the Donner Party.

I digress (which I predict I'll be doing quite a bit in this book). This year, in my senior year of high school, we decided to go white water rafting on the South Fork American River. Going into the trip, I was unfamiliar with white water rafting. In fact, it wasn't until the end of the trip that I realized it's called white water rafting because river rapids appear white and not because whoever named it just liked alliteration.

Speaking of which, have you ever wondered why rapids and waves appear white? After all, water is clear, so what's going on here? When water gets tossed around — as it would in rapids and waves — a bunch of bubbles are created with the help of the surrounding air. Sunlight itself is white because the sun emits every colour of visible light, all of which combine to make white. In contrast with a water droplet (which absorbs some wavelengths of visible light and reflects others), a bubble filled with air doesn't absorb sunlight well at all, and, thus, appears brighter and whiter than its surrounding environment.

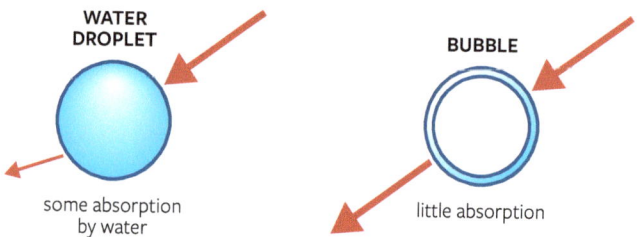

*Fig. 2.* How light interacts with water droplets and bubbles

I didn't know any of this on the Monday my classmates and I took off for the South Fork American River, but I was more than okay with that.

At this point in my life, I've become quite fond of exposing myself to completely new situations; when I find myself in a new place, I not only learn something new about the world, but I also learn something new about myself. Thus, I love jumping into the deep end, something I did both figuratively and literally while rafting.

This brings us back to the time when Luke, the biology teacher from my school who was also in my raft, asked us, "What's the most beautiful thing you've ever seen?" Some people had to think for a second to answer the question, but not me; I could answer it instantly. The most beautiful thing I've ever seen was the night sky at the Mountain School in Vermont.

I remember the first moment I witnessed that night sky. Not a single light in any building was on for miles, and, because of that, I could see a night sky painted with an endless amount of stars. The night sky engulfed me and invited me to explore it with the same curiosity and thirst for the unknown that a young child has when he begins to learn how big the world truly is. And, although I could barely see it in the dark, I was surrounded by a forest of trees, all of which parentally encouraged me to explore on.

It was euphoric.

*Fig. 3.* The night sky at the Mountain School. Yeah, my camera couldn't capture any of the stars but it was breathtakingly beautiful, I promise.

You probably have a lot of questions at this point. What is the Mountain School? What's the point of that story? Is Alex even allowed to address the reader by writing in the second person in a nonfiction book? And, most importantly, how does any of this relate to compost?

Let me explain, and let me start with the first question: what is the Mountain School?

To fully understand what the Mountain School (or TMS, for short) is, we must first understand the history of TMS. And that history starts with a man named Mac Conard.

On April 17, 1925, in Bridgeport, Connecticut, Waller MacNiven Conard — or Mac Conard, for short — was born. He was the son of his dad, Frederick, and his mom, Julia. He had three older brothers.

Mac attended Kingswood School for his pre-collegiate education and graduated in 1943. He attended Yale, but before he finished his education he was called to fight in World War II. He served in the U.S. Navy aboard the Fleet Minesweeper USS Raven and then returned to Yale to graduate in 1948. Seeing as he had some experience in the Navy and quite enjoyed boatbuilding, he started working in the machine tool industry.

Unfortunately, this was not the right choice for him. In the years following 1948, he quickly became dissatisfied with his job. On a more positive note, however, he met and fell in love with Doris Emerson; they got married in 1951. Doris had a background as a secondary school teacher, and Mac realized he found education much more fulfilling than building tools, so he joined Doris and moved to Putney, Vermont where they taught at the Putney School. After five years of that, Mac returned to Yale to get a Master's degree in conservation at the Yale School of Forestry & Environmental Studies. Mac and Doris wanted a way to combine their passion for education and their passion for environmental sustainability, so they decided to start a high school which did just that: the Mountain School.

On May 10th, 1962, they bought a farm in Vershire, Vermont and began building the Mountain School, a four-year boarding school for high schoolers. They built everything from the buildings to the curriculum from scratch, all the while staying true to their original vision of academic excellence and sustainability. The school was very successful, but, twenty years later, Mac and Doris decided to retire. However, they wanted to make sure the Mountain School was in good hands, so they began talking with David and Nancy Grant about the future of the school. They decided to transform the school from a four-year boarding school to a semester-long program for high school juniors. Now the only thing they needed was funding.

David and Nancy were both affiliated with Milton Academy, a boarding school in Milton, Massachusetts, so they decided to ask the Trustees of Milton Academy to purchase the Mountain School. The following is an excerpt from David and Nancy Grant's proposal to the Trustees:

> We stood on the hill in the center of the 300-acre campus and saw the evidence of a remarkable educational opportunity. The huge school garden stretched before us and sloped down to the field where a few sheep and cows grazed beside the barn. At the end of the field, we could look down on the buildings of the school.... Beyond the campus stretched Vermont and New Hampshire, the peaks of the White Mountains visible in the east. We were struck both by the fact that the Mountain School students lived and worked close to the natural world and by our own sense that to appreciate this work and one's part in it is educational in some fundamental way.

Milton Academy purchased the Mountain School in 1983, and the semester program began. If you're wondering what happened to Mac and Doris, don't worry; it turned out very well for them. They had three children — Nathaniel, David, and Peter — and were very active in the Vershire community. Doris passed away in 2009, and Mac passed away on May 10th, 2013, exactly 51 years after he and his wife first arrived in Vershire.

*Fig. 4.* The Mountain School's campus in winter

Let's fast forward to 2018, my sophomore year of high school. Sophomore year was a weird year for me. Around April, I started feeling intense burnout. While I usually love my classes, I was failing to find meaning in the work I was doing. Moreover, I wasn't super close with anyone else at my school; in fact, despite yearning for deep human connection, I began to distance myself from my closest friends. Life began to lose its purpose. Pro-tip: if you ever get to a place in life where, day in and day out, you think, "it doesn't matter what I do today because, in a month, I'm gonna forget that this day ever existed," it's probably a good time to consider changing some things up.

The one time I did feel at peace, however, was when I was biking up and down the San Francisco coastline. I began biking after becoming frustrated with how sedentary my lifestyle was. Between making videos for my educational YouTube channel Technicality (a job which mostly consists of sitting at my computer and either researching, scripting, or editing) and doing homework, too much of my life was spent sitting down. So, one day, after school, I decided to go biking. I followed rivers. I conquered trails. I never had any idea where I'd go when I started biking, but that didn't matter: the joy was in the exploring.

*Fig. 5. A place I went to one day whilst biking*

This quickly became a habit. Every day, when the clock hit 3:30pm and classes let out, I grabbed my bike, put on a Kendrick Lamar album, and got lost for two hours. Even though I started biking for exercise, it evolved into something even greater; I began to develop a fascination with the concept of place. Suddenly, my school wasn't just a place I learned things at for eight hours a day, but part of something bigger: the San Francisco Bay Area environment. By getting lost in it, I gained a richer understanding of the world around me, and, thus, my place within it.

*Fig. 6. Some ducks I saw whilst biking*

One day, a woman named Emily Sartin came to my school. She went to the Mountain School when she was a junior in high school and, upon graduating from Harvard, returned to TMS to work as a graduate resident. Her duties as a graduate resident included, among other things, being an advisor to a handful of students, leading work crews on the farm and in the kitchen, and visiting schools in hopes of telling students about TMS.

That was the day I first learned about the Mountain School. I learned about how the Mountain School combines rigorous academics with farm work and outdoorsmanship. I learned about how, at the Mountain School, people strive to understand and connect with each other on a deep level. And I learned about how one of the Mountain School's core values is "knowing a place."

I had a gut feeling: the Mountain School was where I needed to be. Although I didn't realize it until after the fact, my life at the time paralleled Mac Conard; we both were dissatisfied in our lives and set out to find a way we could be fulfilled in our education and, at the same time, live amongst nature.

*Fig. 7. A panoramic view from the Mountain School's Garden Hill*

And with that, I jumped into the deep end. In August 2018, I moved to rural Vermont to begin my semester.

*Fig. 8. I took this photo of the sunset at the Mountain School two days after I first got there.*

The first couple weeks at the Mountain School were rough. At TMS, a typical Monday would consist of everything from farming to classwork to preparing to go camping. At my home in California, a typical Monday would consist of me sitting at my desk writing a script about whatever cool concept I'm into that month. The Mountain School was a departure from what I knew in the most radical way possible and, although I started TMS with the usual enthusiasm and optimism I approach all new experiences with, those emotions quickly faded as I began to feel more stressed and alone. On top of all of that, there was no one I could talk to about this. I had only been there for two weeks, so I didn't know anyone well enough to feel comfortable talking about it. Plus, even if I did feel comfortable talking with someone, would I even know how to? It's not like I was super great about opening up to people and talking about what I'm going through with them; I never really did that back home.

I came up with a fool-proof strategy for dealing with these emotions: ignore them. After all, since I had always loved jumping into the deep end before, struggling with being in a different environment was super unusual for me. So, instead of actually confronting those emotions, I just decided to suppress them and adopt my usual personality of enthusiasm and optimism instead.

However, everything all came crashing down on September 6th, 2018. Before I went to bed, I thought it would be a good idea for me to charge my computer overnight. But when I checked my backpack for my MacBook, it wasn't there. I suddenly realized what had happened: not only did I lose my computer (which in and of itself is pretty bad), I also lost all of the work I had done on my next Technicality video.

This was the tipping point. I had been separated from my parents, my friends, my home, and now, the machine that allowed me to further my life's work, Technicality. I wrote about that feeling in my journal that night:

*Goddamn everything I suppressed came back up. It finally hit me that I was away from my parents. It finally hit me they couldn't help me anymore.*
*It finally hit me I was alone.*
*So for a solid 40 minutes, I just lay there crying. I couldn't stop. I tried to stop but I just couldn't.*

I had thrown myself into the deep end, and, for the first time in my life, instead of learning to swim, I was drowning.

Realizing I had no option but to fix this quickly, I finally tried something I never thought I could do: ask for help. Even though I didn't know many of the other people at TMS too well, I began talking with some of my newly-made friends about how I was feeling. I told them I thought I was in over my head and I missed home and I had no idea what I was doing. Turns out, many (if not all) of them said they were feeling the exact same way. They understood what I was going through, and I understood what they were going through. I wasn't alone anymore. Together, my friends and I began to learn how to swim.

*Fig. 9. A moment on Garden Hill*

Things started to turn around after that. I had finally found the deep human connection I was searching for back in California, and after that other things seemed to fall into place as well. For instance, I began to find a lot of meaning in my work. I was taking five classes — English, History, Calculus, Environmental Science, and Humanities — and was fascinated by the things I was learning. Outside of the classroom, I took a liking to the Mountain School's apple orchard. We had to do two hours of farm work daily, and I would frequently work with the head of the apple orchard, Ben Tiefenthaler, on whatever tasks needed to be done for the day, whether they be apple picking or moving the electric fence or making cider.

*Fig. 10.* Desmond, TMS's Guard Llama, eating an apple from a tree in the orchard.

*Fig. 11.* Wait, did Alex really just write "Guard Llama"? Yup, we had a guard llama. His name was Desmond. We all liked him very much.

On top of the two hours of farm work daily, we also attended a 50-minute long class called "Farm Seminar" every Wednesday. Every week in Farm Seminar, we learned about a different aspect of agriculture and our food system from various farm staff. For instance, when Ben taught Farm Seminar, we learned about the history of apples in the United States and the Newtown Pippin.

The Newtown Pippin is the most important apple you've never heard of.

*Fig. 12.* A drawing of the Newtown Pippin

When I first ate a Newtown Pippin, three things stood out to me. First, the apple was a bit smaller than I was expecting. While the apples I was used to — such as Galas, Fujis, or Granny Smiths — were generally around the size of my palm, the Newtown Pippin was noticeably smaller than that. Second, the Newtown Pippin had amazing crunch. The apple was strong, sharp, and firm, and biting into it was incredibly satisfying. And third, the apple had stupendous flavour, as perfectly described by pomologist David Karp:

> *Its clean, highly aromatic flavor, a blend of pine, citrus and walnut that somehow epitomizes apple, has an ideal balance of sweetness and acidity and a complex, lingering aftertaste.*

The apple tastes so good, it actually played a role in eighteenth and nineteenth century English politics. But to understand that, we first have to understand the origin of the Newtown Pippin.

*Fig. 13. A photograph of the Newtown Pippin*

Gershom Moore was the first person to discover the Newtown Pippin in 1720. Moore owned an apple orchard in Newtown, New York. One day, he ate an apple from one of his apple trees and realized that it was far more delicious than any apple he had ever eaten before. He quickly named the apple after the city it was grown in and began to sell scions of the apple tree. What are scions?

Apples are pretty sneaky fruits. If you plant a seed from a certain apple, you will not get an apple tree that produces the same variety of apple as the one you acquired the seed from; you'll get a completely new variety of apple. Thus, the only way to reproduce a variety of apples is through a process called grafting, which involves combining the tissue of two plants so those plants can grow together. The plant tissue used in the lower part of the grafted plant is called the rootstock, and the plant tissue used in the upper part of the grafted plant — the part that produces fruit — is called the scion. By merging a scion from the Newtown Pippin and a rootstock from some other plant, farmers could create a Newtown Pippin tree and reproduce the apple.

The apple quickly swept the nation, which was especially impressive considering it was the 18th Century and the only things that swept the nation back then were stuff like the First Great Awakening and smallpox.

Horticulturist William Coxe called the Newtown Pippin "The finest apple of our country, and probably of the world." George Washington grew Newtown Pippins at Mount Vernon. Thomas Jefferson grew Newtown Pippins at Monticello; in fact, at 170 Newtown Pippin trees, he had more Newtown Pippins than any other variety of apples. When he lived in Paris serving as America's ambassador to France (or Minister to France as it was called at the time), he wrote a letter to James Madison in which he complained, "[the French] have no apples here to compare with our Newtown Pippin."

Benjamin Franklin was introduced to the Newtown Pippin when he was in London in 1759. Naturally, Franklin loved the apple and began sharing it with his friends, all of whom loved it as well and wanted more. England tried to import a scion of the Newtown Pippin, but its climate proved challenging for the pippin to grow, so people had to resort to importing individual Newtown Pippins from America. Unsurprisingly, this made the apples very expensive, and they became a symbol of wealth and decadence.

This whole event is known as the First Pippin Mania in England. Yes, you read that correctly: the *First* Pippin Mania. The Second Pippin Mania happened in 1838. England had recently implemented a tariff on imported apples in the hopes of catalyzing the small apple industry they did have. However, when Andrew Stevenson (America's ambassador to Great Britain at the time) gave a basket of Newtown Pippins to Queen Victoria as a gift, Queen Victoria — then only eighteen years old — immediately lifted the tariff upon tasting the apples. Andrew Stevenson's wife, Sally Coles Stevenson, later wrote that the Newtown Pippins were, "...eaten and praised by royal lips and swallowed by many aristocratic throats."

Shortly after, Newtown Pippins became popular once again, selling for three times more than the average apple. People even had to be cautious of fake Newtown Pippins; some shady sellers would call similar-looking apples Newtown Pippins and sell them for a lot of money at markets in London. The Newtown Pippin was the Supreme of apples.

Sadly, in the words of Robert Frost, nothing gold can stay. During the Great Depression, many Newtown Pippin growers couldn't make ends meet, and the Newtown Pippin faded from America's east coast.

However, it's not a sad ending for our beloved apple. Before the Depression, in the 1850s, the Newtown Pippin began to make its way out west. Newtown Pippin trees popped up in Watsonville, a city that's about 90 miles southeast of San Francisco. The weather in Watsonville is perfect for Newtown Pippins; Watsonville has a long growing season, a moderate climate, and plenty of coastal fog which protects the apples from intense sun. By 1920, 12,000 acres of Newtown Pippins dominated Watsonville, and many immigrants from Dalmatia (a region in Croatia) moved to Watsonville to become apple farmers.

Today, you won't find any Newtown Pippins in a grocery store. Despite their superior taste, they don't look as appealing as a Granny Smith apple, which is bigger and (for the most part) has fewer imperfections, so it's widely preferred by both produce sellers and consumers. Nonetheless, the Newtown Pippin still lives on. Notably, Martinelli's (a popular high-end apple juice and cider company) buys 85% of Newtown Pippins grown in the Watsonville region, and about half of Martinelli's juice is made up of Newtown Pippins. Martinelli's pays over $200 a ton for Newtown Pippins — significantly higher than the going rate for juice apples — in order to ensure farmers don't replace their Newtown Pippin trees with trees that are more profitable.

*Fig. 14. Martinelli's apple juice*

So that's a taste of what one Farm Seminar period would be like. Needless to say, Ben's Farm Seminar was one of my two favourite Farm Seminar lectures; the story of the Newtown Pippin fascinates me to this day. My other favourite Farm Seminar lecture was on compost. It was led by Sam Kelman, the person in charge of both the compost program and the sugaring program, which turned maple tree sap into maple syrup.

Food waste was dealt with superbly at the Mountain School. After every meal, we would scrape all of our food waste into a bucket which was then thrown into one of the four massive piles of compost we had. We could then use that compost on our farm to grow our food, some of which we ate and some of which became food waste, allowing the cycle to continue and continue.

In this Farm Seminar lecture, Sam discussed not only the logistics of compost at the Mountain School, but also the logistics of compost in urban areas, such as New York City. I was instantly captivated by this; I found immense pleasure in the concept of harmony between what we eat, what we throw away, and what food we grow. Thus, I set out to learn everything I could about composting and specifically — considering that ~80% of Americans live in urban areas — how compost programs could be implemented in cities. I visited my local compost facility, completed a research project on compost for my environmental science class, and produced a short-form documentary on urban composting, which I uploaded to my aforementioned YouTube channel, Technicality.

In the modern age, we're incredibly disconnected from our trash. Every night after dinner, I throw my food waste away in the trash and don't even think twice about it, and (if I were of legal age to be a betting man) I'd bet you do something similar too.

I want to change that disconnectedness, and the first step to doing that is understanding the breadth of the problem which stands before us.

# Part 2:
# Why We Need City-Wide Compost Programs

*Or, everything you need to know about
food waste and climate change.*

The United States has a massive food waste problem. About 40% of food in the US — or around 1,500 calories per person per day — goes uneaten. That's over 87 million Big Macs worth of food calories wasted daily, which is over two times as much as most other industrialized nations, and enough calories to feed the entirety of Germany, Canada, and Australia combined (around 150 million people). Moreover, the amount of food we've been wasting has been increasing; food loss today is 50% more than what was lost in the 1970s.

Breaking food loss down into more specific categories: we waste around half of all fruits, vegetables, and seafood, one-third of all grains, and one-fifth of all meat and dairy.

Ah, but Alex, you might say, I read some study once that said just 100 companies are responsible for over 70% of global $CO_2$ emissions, so while it is helpful that I take the train instead of driving, and use my Hydro Flask™ instead of a plastic bottle, if we really want to curb climate change, we need to have top-down action. And wouldn't it be the same for food? I mean yeah, I don't finish my kale sometimes but (1) can you blame me and (2) that can't be the real cause of the problem; it must be big food up to their old, not-green habits.

Nope! While it is indeed correct that 100 companies are behind over 70% of $CO_2$ emissions, the trend of companies and not consumers being largely at fault does not continue with food. 17% of food loss occurs on the farm while harvesting the food, 6% occurs in handling and storage,

9% in processing, 7% in retail, and 61% in consumption. If we, as consumers, worked to eliminate food waste, we'd make some very considerable progress in solving our food waste problem. And solving food waste would be a pretty impactful thing to solve.

30% of our fertilizer use, 31% of cropland use, 25% of total freshwater consumption, and 2% of total energy consumption goes into producing food we don't even eat.

Food we don't eat ends up in landfills; as a matter of fact, there's more food in landfills than any other thing we throw away. To understand the true breadth of the impact that food waste has in landfills, we first have to learn about the history of Earth's global climate and anthropogenic climate change.

*Fig. 15.* "Earthrise," a famous 1968 photograph of the Earth from the Moon taken by Apollo 8 astronaut William Anders.

It's probably a good idea to start by defining what climate is. Climate describes the long-term average weather trends of a certain area. Keep in mind: climate is not the same thing as weather. While weather describes the atmospheric conditions in one specific place and time, climate describes the broader patterns and changes of weather over time.

What has been Earth's global climate for the past 4.5 billion years? In a word, habitable. We know that the Earth has been consistently habitable for at least the past 3.7 billion years and possibly longer because we have fossil evidence of life from billions of years ago.

*Fig. 16. A 1.4 billion-year-old fossil of Horodyskia*

Moreover, we also have evidence that liquid water has been on Earth for the past four billion years because we found sedimentary rock deposits formed from rocks affected by water that are that old. Evidence of liquid water means that Earth's temperature must have stayed between 32°F (0°C) and 212°F (100°C), or somewhere between freezing and boiling. This is certainly true right now; the average temperature of the Earth is 59°F (15°C), and a large majority of our surface is between 32°F (0°C) and 86°F (30°C), which is perfect for life.

There are two different aspects involved in changing the Earth's climate: climate forcing and climate response. Put simply, a climate forcing is a factor which *causes* Earth's climate to change, while a climate response is an *effect* of a climate forcing.

There are two different types of climate forcings: natural and anthropogenic. We'll get to anthropogenic climate forcings in a bit, but first let's learn about three of the most notable natural climate forcings.

## CLIMATE FORCING FACTOR ONE: PLATE TECTONICS

First, plate tectonics, which operate over the span of millions of years, affect Earth's climate. The theory of plate tectonics states that the Earth's crust is not solid and stagnant, but made of various plates and constantly moving. You can think of Earth's surface like the shell of an egg: it's weak and easily broken.

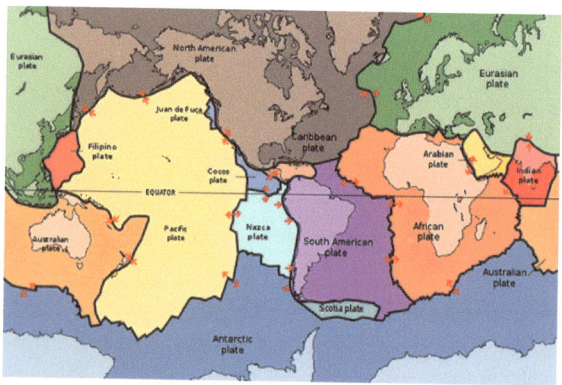

*Fig. 17. Earth's plates*

Earth's crust is all broken up like this because of the forces inside of the Earth. Factors like heat and gravity break up Earth's crust into many smaller pieces.

While Earth has dozens of plates, the five largest are...
- Pacific Plate (103,300,000 km²)
- North American Plate (75,900,000 km²)
- Eurasian Plate (67,800,000 km²)
- African Plate (61,300,000 km²)
- Antarctic Plate (60,900,000 km²)

There are two types of plates: oceanic and continental plates. Oceanic plates are mostly made of rocks like basalt, while continental plates are made up of rocks like granite. As a result, continental plates are less dense and float above oceanic plates.

Earth's plates move because of something known as convection currents. What are convection currents?

Particles in a certain substance are always moving. The faster the particles in a certain substance are moving, the hotter the substance is, and the slower the particles in a certain substance are moving, the colder the substance is. Indeed, temperature is defined as the average kinetic energy in a certain system, or, in other words, how quickly the particles in a substance are moving.

You know how hot air rises? Well, that's because when air gets hot, the molecules which make up the air move faster and spread out, thus making the hot air less dense than the surrounding air. As a result, the less dense air rises to the top and the denser air drops to the bottom.

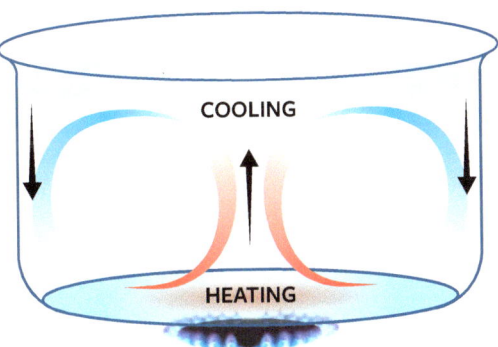

*Fig. 18. The hot fluid rises and the cold fluid sinks*

All fluids act like this, and the magma in Earth's mantle is no exception. Magma is incredibly hot semi-liquid rock that's found in the Earth's mantle. Due to the heat of the Earth's core, the magma deep in Earth's mantle is significantly hotter than the magma closer to Earth's surface.

Since hot fluids rise, the hot magma deep in Earth's mantle will rise up and the relatively cooler magma close to the Earth's surface will drop down. *That* magma then gets heated by the Earth's core, causing it to rise and the magma near the surface to sink. This very process continues and continues, albeit pretty slowly because magma is *super* viscous. This process is known as a convection current and is what moves the plates around the Earth.

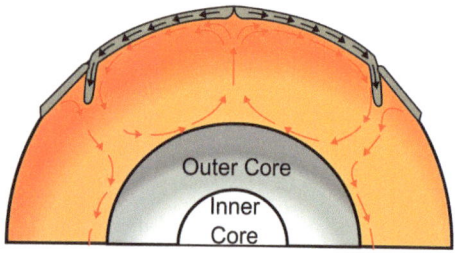

*Fig. 19. How convection currents work*

The moving of Earth's plates inevitably leads to plates colliding, separating, and interacting. There are three different kinds of plate boundaries, and each one describes a different way that various plates interact. First, at divergent plate boundaries, the two tectonic plates involved move away from each other. In return, magma from the uppermost part of the mantle rises up and fills the newly created gap. The magma then cools, becomes solid, and creates new land. This is commonly seen in the ocean, specifically when creating ridges (like the Mid-Atlantic Ridge). However, it can also occur between two continental plates, which is the case for the East African Rift Valley.

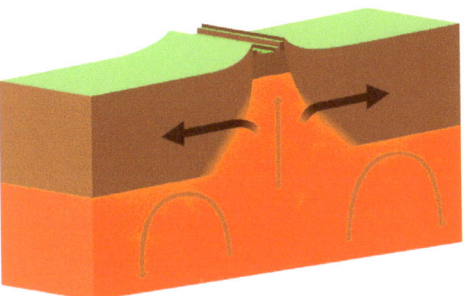

*Fig. 20. Divergent plate boundary*

Second, at transform boundaries, the two tectonic plates involved slide against each other. I, a Californian, am particularly familiar with transform boundaries because the movement of plates at a transform boundary results in an earthquake. Examples of transform boundaries include New Zealand's Alpine Fault and California's San Andreas Fault.

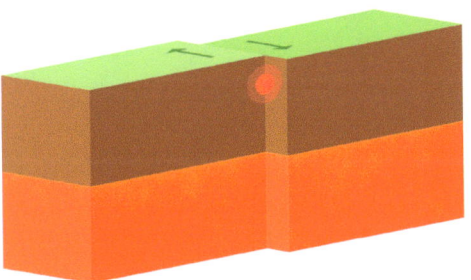

*Fig. 21. Transform plate boundary*

Third, at convergent plate boundaries, the two tectonic plates involved move towards each other. The denser plate (which will be the oceanic plate if the convergent plate boundary is between an oceanic plate and a continental plate) goes under the less dense plate. This is how the islands of Japan and mountain ranges like the Himalayas or the Andes were formed.

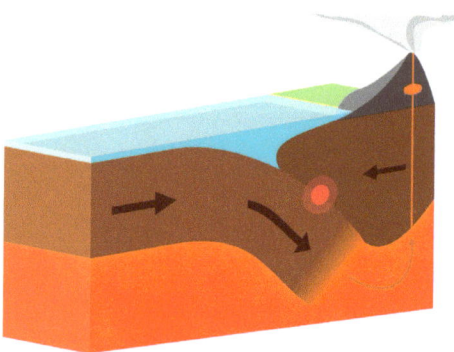

*Fig. 22. Convergent plate boundary*

The origin of plate tectonics can be traced back to the work of Alfred Wegener in the twentieth century. Wegener hypothesized that the continents moved, a concept he called (perhaps unsurprisingly) continental drift.

*Fig. 23. Alfred Wegener*

Wegener cited a variety of observations as evidence for his theory. First, Wegener noticed that all of the continents look like they fit together. If you treated each continent as a piece in a jigsaw puzzle, you could rearrange them and they would all fit together nicely.

Second, Wegener noticed that not only did the outlines of the continents match, but so did the geology. Rocks like gypsum are only created in a desert environment, but he found gypsum in places that weren't traditionally desert. Likewise, coal — which only forms near the equator — was found in places far from the equator, and evidence of glaciers (such as glaciers shaping and molding rocks) was found in places that don't typically have glaciers.

He theorized that there must have been a supercontinent, Pangea, hundreds of millions of years ago where all of these things were created, and since then the continents drifted to where they are today.

*Fig. 24. Pangea*

Prior to Wegener, the previously held notion was that Earth's crust is stagnant, so the scientific community received Wegener's theory quite negatively at the time. Among other things, they complained that Wegener started with a hypothesis and looked for evidence to support that hypothesis as opposed to looking for evidence first, then generating various hypotheses that explain the evidence, and finally choosing the best hypothesis.

The idea lay dormant for around twenty years after Wegener's death in 1942 until geologist Henry Hess researched a concept called seafloor spreading.

Seafloor spreading states that oceanic plates are constantly moving, and magma from the mantle is constantly moving up to Earth's surface and creating new land.

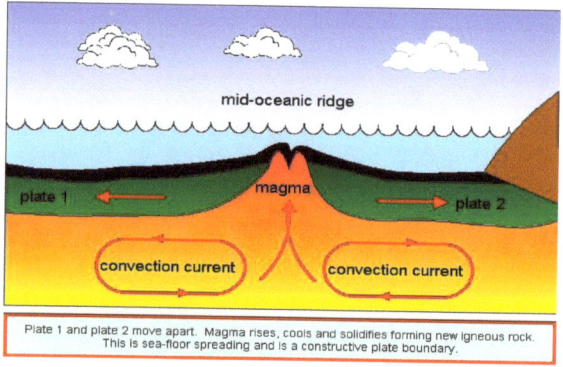

*Fig. 25. Seafloor spreading*

However, we've now got a problem: if oceanic plates are constantly moving and the cracks are being filled by the magma in the mantle, where do those oceanic plates go? It's not like the crust is getting bigger. Hess found that the oceanic plates will sink back down into the mantle, and — every couple hundred million years — the entirety of the ocean's crust is recycled into itself. Hess was so pleased by this revelation that he also called it geopoetry.

In 1963, scientists made an interesting observation which helped prove Hess's hypothesis: they found that the crust on the ocean floor is divided into 'stripes' that alternate magnetic polarity.

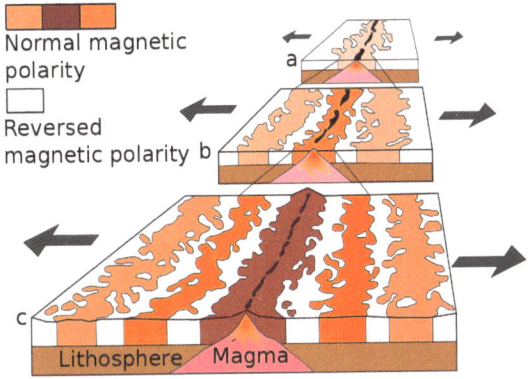

*Fig. 26. This is a diagram of magnetic stripes on the ocean floor; Ridge A illustrates the seafloor 5 million years ago, Ridge B illustrates the seafloor 2 to 3 million years ago, and Ridge C illustrates the seafloor in the present day.*

What's going on here? Well, certain rocks have magnetic iron crystals in them. These certain rocks are formed by liquid magma cooling and solidifying. However, before the magma cools down, the magnetic iron crystals are free to move around within the liquid and, thus, orient themselves to magnetic north. If a rock has magnetic iron crystals in it, those crystals will point towards magnetic north when the rock is being formed. Then, when the rock solidifies, those crystals get locked into that orientation. Magnetic north has shifted many times throughout the Earth's history, so these crystals can tell scientists a lot about the orientation of various chunks of rock at different points in time. Indeed, scientists found that rocks from North America and Europe are oriented in a way that supports Hess's hypothesis.

Possible climate responses to changes in plate tectonics can be changes in factors like temperature, precipitation, snow cover, vegetation, wind, and the circulation of ocean waters.

## CLIMATE FORCING FACTOR TWO: ORBITAL CHANGES

Second, orbital changes of the Earth — which affect the amount of sunlight the Earth receives — affects Earth's climate. Earth's orbit is not always constant; instead, different aspects of Earth's orbit come in cycles known as Milankovitch cycles. Milankovitch cycles are the different ways the Earth moves around the sun. There are three different Milankovitch cycles: eccentricity, precession, and tilt.

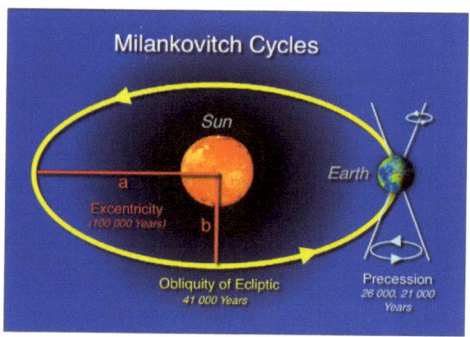

*Fig. 27. Milankovitch cycles*

The first Milankovitch cycle, eccentricity, is a measure of how elliptical Earth's orbit is. To understand eccentricity, two misconceptions must be debunked. The first misconception is that Earth's orbit is a perfect circle. As a matter of fact, it's more of an ellipsoid than a circle. Eccentricity describes how stretched out that ellipse is. The second misconception is that the Sun is at the center of the Earth's orbit; the Earth is actually three million miles closer to the sun on one side of its orbit than on the other. The point where the Earth is farthest away from the sun is called the aphelion and the point where it's closest is the perihelion. These points are opposite to each other.

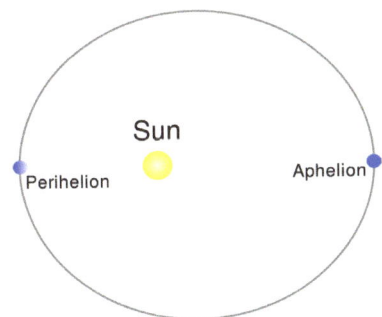

**Fig. 28.** *Eccentricity (not to scale)*

Eccentricity changes every 100,000 years. That means that every 100,000 years, the orbit of the Earth stretches out and compresses back in. Eccentricity was first discovered by French mathematician and astronomer Urbain Le Verrier, and it combines with precession to affect the Earth's distance from the sun and, thus, the intensity of Earth's seasons.

**Fig. 29.** *Urbain Le Verrier*

Much like a top as it slows down, the Earth wobbles. This wobble lasts 22,000 years; at year 0, the Earth is tilted to the star Vega, at year 11,000, the Earth is tilted to the star Polaris, and, at year 22,000, the Earth is back at Vega. Precession — the second Milankovitch cycle — is a measure of where the Earth is in its wobble.

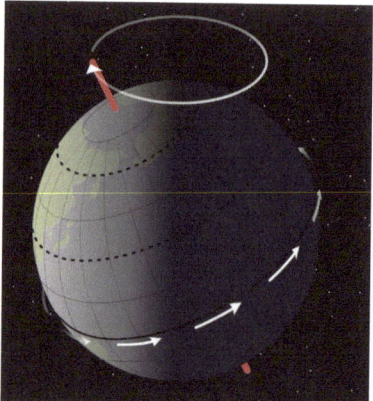

*Fig. 30. Precession*

Currently, in the 21st century, the Earth is pointed towards Polaris. Thus, Polaris is also known as the North Star. The first person to discover precession was Jean le Rond d'Alembert, a French mathematician.

*Fig. 31. Jean le Rond d'Alembert*

Eccentricity exacerbates precession; when eccentricity is higher, precession cycles are larger in amplitude, and when eccentricity is lower, precession cycles are smaller in amplitude.

The combination of eccentricity and precession affect Earth's distance from the sun. And, of course, Earth's distance from the sun affects how much the sun's heat affects the Earth.

The third Milankovitch cycle, tilt, measures (you guessed it) how much the Earth tilts. Earth's axis is not perfectly vertical but rather tilted on its side. Tilt is the source of Earth's seasons; summer is when the Northern Hemisphere is tilted to the sun, and winter is when it's tilted away from it.

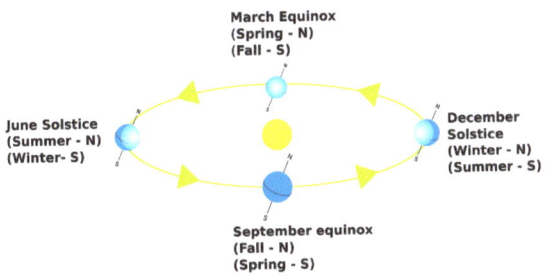

*Fig. 32. Earth's axis remains pointed in the same direction throughout the Earth's orbit*

Seeing that it is one of the Milankovitch *cycles*, Earth's tilt is not constant. The length of tilt's cycle is around 41,000 years. While the Earth's axis is currently tilted at 23.5°, the less extreme end of the tilt cycle is 22.1° and the more extreme is 24.5°. At 22.1°, sunlight is spread out in a way that produces more moderate seasons, and, at 24.5°, sunlight is spread out in a way that produces more extreme seasons.

34 • Behold This Compost

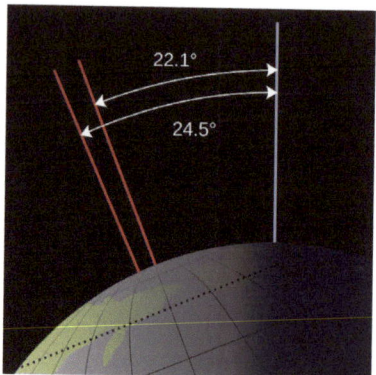

*Fig. 33. Tilt*

The combination of tilt, eccentricity, and precession can be used to predict the intensity of Earth's seasons. Serbian mathematician and astronomer Milutin Milankovitch, despite not being the first person to discover each of the three Milankovitch cycles, did discover that the combination of all of them affected the intensity of our seasons. Thus, the cycles are named after him.

### CLIMATE FORCING FACTOR THREE: THE STRENGTH OF THE SUN

Third, the strength of the Sun has slowly increased over the past 4.6 billion years as a result of an increase in hydrogen fusion in the Sun's core. It's important to note that this increase isn't insignificant; the sun now shines 25%-30% brighter than it did billions of years ago, which affects Earth's climate.

*Fig. 34. The sun from NASA's Solar Dynamics Observatory*

As the sun gets stronger and stronger, the Earth receives more and more sunlight. This, of course, affects Earth's temperature, and that affects many aspects of Earth's climate, such as snow cover, precipitation, wind, etc.

So now that we have an understanding of some of the most prominent *natural* climate forcings, let's learn a bit about man-made greenhouse gasses, an *anthropogenic* climate forcing.

First, we know something seems to be wrong. In recent history, the temperature of our globe is rising disproportionately to the strength of the sun.

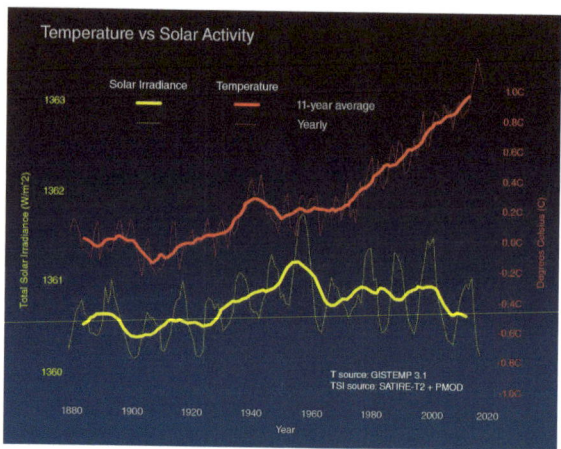

*Fig. 35. Total solar irradiance in watts per meter squared and temperature change in degrees celsius versus time in years*

What's going on? Greenhouse gases — or GHGs, for short — are any gasses in the atmosphere which absorb and emit heat. Examples include carbon dioxide ($CO_2$), methane ($CH_4$), nitrous oxide ($N_2O$), and various other gasses.

From around 10,000 years ago up to the Industrial Revolution, the amount of greenhouse gasses in the Earth's atmosphere stayed pretty constant. However, since then, we've seen a 40% rise in carbon dioxide and a 150% rise in methane. Moreover, annual anthropogenic GHGs increased by 81% since 1970.

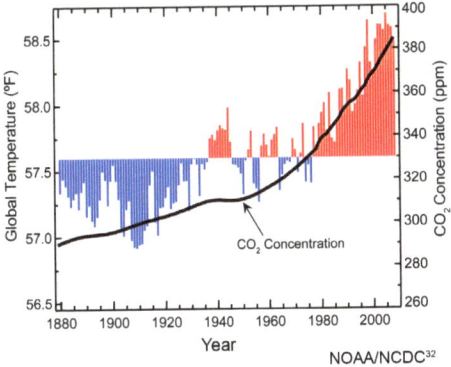

*Fig. 36.* CO2 concentration in parts per million and global temperature in degrees fahrenheit versus time in years

Take carbon dioxide, for example. The $CO_2$ we emit mainly comes from the burning of fossil fuels. In fact, as can be seen in the diagram of the carbon cycle below, 5,000-10,000 petagrams of carbon are stored in fossil fuel deposits.

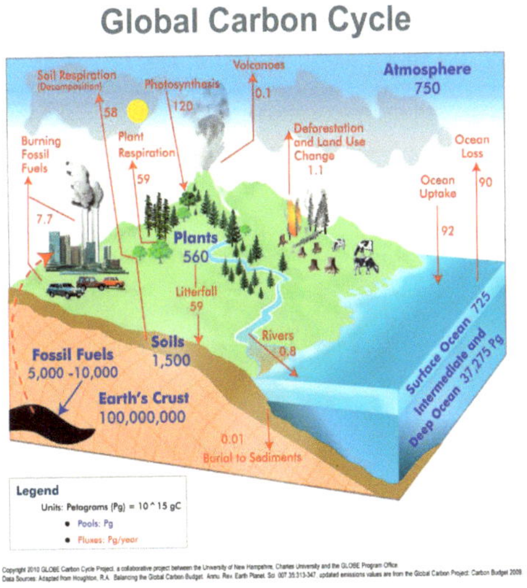

*Fig. 37.* The carbon cycle

Fossil fuels are called fossil fuels because they come from organic matter. Petroleum is a liquid fossil fuel and coal is a solid.

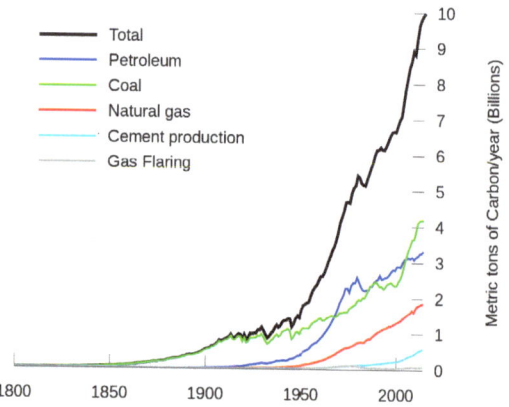

*Fig. 38.* Global carbon emissions in billions of metric tons per year versus time in years

The burning of fossil fuels currently provides America with 80% of its power, but, while it does that, it releases the carbon stored in it into the atmosphere. Figure 39 illuminates what we're doing with the fossil fuels we're burning.

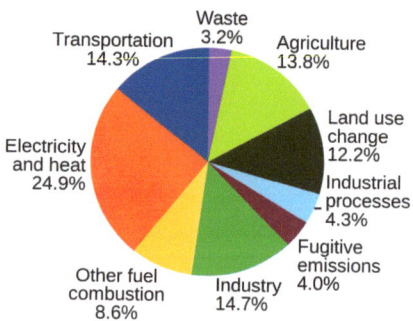

*Fig. 39.* World greenhouse gas emission in 2005 by sector

But, of course, $CO_2$ isn't the only GHG that's surging in concentration in our atmosphere. This is where food waste comes back into the picture: when food waste is put in landfills, it produces Landfill Gas (or LFG), which is about half carbon dioxide and half methane. Some modern landfills have methods of capturing LFG and either destroying it or turning it into fuel or energy, but that's certainly not the majority because landfills still account for over 14% of methane emissions in America in 2017, making them the third-largest source of man-made methane emissions in the US.

All in all, the amount of global food loss in 2009 alone was the cause behind 3,300–5,600 million metric tons of greenhouse gas emissions, and these greenhouse gas emissions have significant negative impacts. To learn about what they are, let's take a look at how the climate has responded to this climate forcing.

Because of the increase in GHGs, we're seeing record high temperatures. The average surface temperature has risen by around 1.4°F (.78°C) since 1800, and two-thirds of that warming has occurred in the last 45 years. 2016 was the hottest year on record, as shown by figure 40, and the temperature is increasing all around the globe, as shown by figure 41.

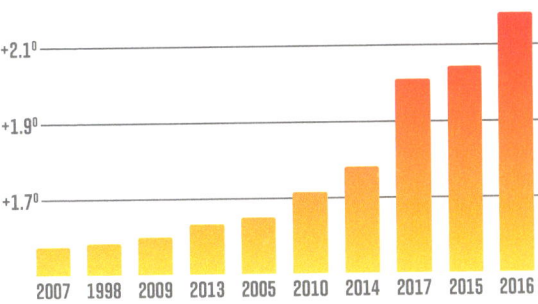

*Fig. 40. The ten hottest years globally in degrees Fahrenheit*

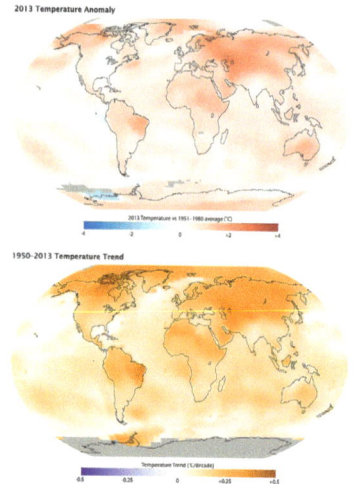

*Fig. 41. A map of temperature in 2013 in degrees celsius versus the average temperature from 1951-1980 in degrees celsius*

This rise in temperature has a domino effect which impacts many other things as well. As seen in figure 42, water is really great at storing heat, so a lot of this new heat is going into the ocean.

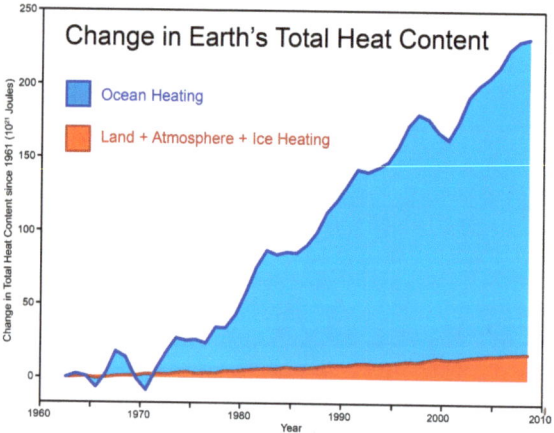

*Fig. 42. Change in total heat content since 1961 in sextillions of joules versus time in years*

Oceans getting hotter causes the sea level to rise because of two reasons. First, water is a fluid, and (much like air and magma), as ocean water gets warmer, it expands. Second, as the temperature increases, more glaciers melt, and the sea level rises. Indeed, the global sea level rose between 6.7 inches (17 centimeters) and 8.3 inches (21 centimeters) in the 20th century.

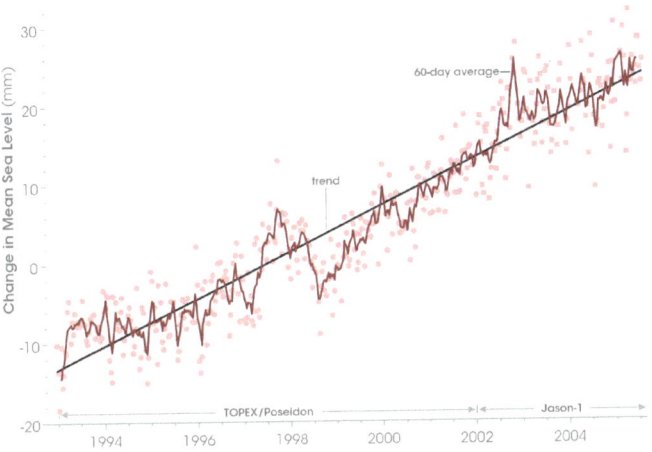

**Fig. 43.** *Change in mean sea level in millimeters as measured during the TOPEX/Poseidon and Jason-1 missions versus time in years*

That's another climate response: arctic sea ice thickness decreased by 40% in the 20th century. Glacial melting is a positive feedback loop; the more glacial ice that melts, the less glacial ice exists to reflect sunlight, and even more glacial ice melts. Figure 44 depicts how much the Easton Glacier on Mount Baker in Washington has retreated since 1985. Notice how much the glacier has shrunk in just 18 years?

*Fig. 44. The retreat of the Easton Glacier*

Moreover, heavy precipitation has greatly increased (20% more rainfall occurs in the heaviest events), and global snow cover has decreased (between 1967 and 2012, snow cover has decreased over 50% in June).

Even if you didn't care about the environmental impacts of food waste or climate change, food waste is bad for your wallet. The average family of four in the US spends around $1,600 a year on food that goes in the trash.

Wow. Okay, that was a lot of information. Here's a picture of a cute puppy to help you digest all of the info you just took in:

*Fig. 45. It was very fun to Google "cute puppy" to get this image.*

So how can we solve the problem of food waste? Well, there are two strategies you can use.

The first strategy you can use is to try to mitigate your personal food waste. I'm sure at one point your parents told you to "take only what you can eat," so follow that and tell others to as well. Moreover, don't treat a food's sell-by date as the last day you could possibly eat that food. One 2015 survey of over 2,200 Americans who considered themselves regular grocery shoppers found that 83% of them had prematurely thrown out food because of that sell-by date, which is quite unfortunate because the sell-by date is merely a recommendation to retailers outlining when they should turnover their inventory.

The second strategy you can use is to try to mitigate the effect of the (hopefully minimal) food waste you do have. Not only can you support initiatives to implement technology in landfills that captures LFG, but also, you can compost.

What is compost? Put simply, it's decomposed organic matter which can be used as a fertilizer. After collecting all of your food scraps, bacteria, fungi, worms, and a bunch of other living creatures get to work at decomposing the former living things and turning them into this rich, black compost.

**Fig. 46.** *Compost*

But, Alex, you might be thinking, if I create a compost pile, it's not gonna help anything. Like, I know the Environmental Protection Agency says composting leads to higher crop yields, but it's not like I'm going around yielding crops all the time in need of compost to better the process. There's nothing *I* can do with a ton of compost.

Well, you're right. Not everyone's a farmer, and not everyone needs compost. So for us to really harness the benefits of compost, we'd need to implement city-wide programs to collect organic waste and turn it into compost for farmers. But, is that even possible?

# Part 3:
# How City-Wide Compost Programs Work

*Or, what I learned when I visited my local composting facility.*

It's a Monday morning. My dad and I are 30 minutes into our 45 minute drive to South Valley Organics, a compost processing plant in Gilroy, California. I take a deep breath in and instantly know we're close. How? Because Gilroy smells like garlic.

Let's rewind to 24 hours earlier. I had completed the first draft of the script for the short-form documentary I was making on urban composting, but I had no idea how I was going to start it. I played around with a couple openers where I would walk around my neighborhood and film other people's organic waste bins, but I wasn't too excited about that; it wasn't super captivating.

Suddenly, however, I got an idea. What if I could actually go to a compost processing facility and film an intro there? I Googled nearby composting facilities and found that the closest one, South Valley Organics, was forty-five minutes away. Immediately, I was disheartened; I don't drive, and I didn't think my parents would (on the spur of the moment) drive an hour and a half round trip so I could see how my organic waste becomes compost.

However, much to my surprise, when I pitched the idea to my dad, he loved it! So the next morning, we got in his car and drove to Gilroy.

At South Valley Organics, around 40,000 tons of organic waste from people all over the Bay Area gets turned into compost. svo actually used to be a landfill, but, around 10 years ago, they covered it up and now it turns organic waste into compost.

And South Valley Organics isn't the only place that does this. Up north, in Vacaville, California, Jepson Prairie Organics processes around 100,000 tons of organic material from Bay Area residents, making it one of the largest food scrap composting facilities in America. South Valley Organics processes organic waste from the South Bay Area (areas such as San Martin, Morgan Hill, and Gilroy), and Jepson Prairie Organics processes organic waste from the rest of the Bay Area (areas such as San Mateo County, Solano County, and San Francisco proper).

As my dad drove down Highway 101, I continued to put finishing touches on my script. Then, suddenly, I could smell garlic. I knew we had entered Gilroy.

Many farmers in Gilroy grow garlic, and the city is famous for its yearly "Gilroy Garlic Festival." Considering that so much garlic is grown there that you can smell it the moment you enter the city, it's no wonder that Gilroy is nicknamed the garlic capital of the world.

After we entered Gilroy, my dad drove a bit further until Google Maps told us to make a left. At first, we were confused; we couldn't see any road to turn into. However, upon looking a bit closer, we realized there was indeed a road, but it was just extremely narrow and made of dirt.

My dad carefully manoeuvred up the dirt road until we got to an office. There, we got out of our car, a woman came out of the office, and I explained who I was:

*Hi, I'm Alex Nickel, I'm a high schooler, I make science documentaries for my YouTube channel, and I'm really interested in compost.*

She told us to wait a second so she could call her manager, and we went back to the car. The office is on top of a big hill, so we had a beautiful view of Gilroy to look at while we were waiting.

*Fig. 47. Our view of Gilroy*

A couple minutes later, the manager drove up in his white pickup truck. I explained who I was again, and he said he was surprised that we were actually interested in the facility and not just lost, since most of the people who drive up to the office are just people who got turned around.

He generously offered to give me a tour of the facility. Eager to learn more, I hopped in his truck, and we drove up to explore the compost piles.

*Fig. 48. The compost piles*

He told me that there are two essential components to a successful city-wide compost program: collection and processing.

## COMPONENT 1: COLLECTION

Collection actually begins far before any garbage truck is deployed, because the first step to collection is making sure residents will actually participate.

To do this, city officials will look towards how many residents are currently composting without a city-wide program put in place. This is exactly what Kathryn Garcia, the commissioner of New York City's Department of Sanitation, did in 2017 to get New York City's composting program off the ground. She pointed out that, in 2016, around 23,000 tons of organic materials were collected from 300,000 households, 722 schools and institutions, and 80 drop-off points.

When cities do propose compost programs, many non-composting residents tend to have concerns like, "What about the smell of food in a food waste bucket?" or "Will that bucket attract rodents?" Fear not, residents, because pilot programs have found that the smell is minimal (if not nonexistent), and it doesn't attract rodents.

Moreover, the key to good organic waste collection is to use previously instituted waste-management infrastructure, such as garbage truck routes and trash-pickup days that have already been established. Not utilizing this infrastructure means your compost program is doomed to fail, as New York City found out around three and a half decades ago when they tested a pilot city-wide composting program and found they didn't have the right trucks or a compost facility that was in a convenient location.

But do you want to know what the biggest hurdle to jump over is when collecting organic waste? It's simply people not knowing what can and can't be composted.

People are pretty terrible at this. Compost facilities have found basically anything you can think of mixed in with organic waste, from pencils to computers. Indeed, the day I visited South Valley Organics, they just got in a new shipment of organic waste, and in it was a plastic helmet and this brick.

***Fig. 49.*** *A brick*

Putting a brick in your organic waste bin would be like putting a picture of a cow in this book; I guess I could understand why you might think it would make sense to do, but it's really out of place and just doesn't belong here.

***Fig. 50.*** *Why is this cow here? This cow doesn't belong here.*

Perhaps unsurprisingly, food waste which comes from schools tends to be especially filled with non-compostable trash. If you're curious as to what can and cannot be composted, please refer to this chart:

| ✓ ACCEPTABLE COMPOSTING MATERIALS | | ✗ UNACCEPTABLE MATERIALS |
|---|---|---|
| **FOOD SCRAPS:**<br>• All fruits and vegetables (including pits and shells)<br>• Coffee grounds and tea leaves<br>• Dairy products (no liquids)<br>• Eggshells and eggs<br>• Leftovers and spoiled food<br>• Cooked meat (including bones)<br>• Seafood (including shellfish)<br><br>**DIRTY, SOILED PAPER:**<br>• Greasy pizza boxes and paper bags<br>• Paper coffee filters and tea bags<br>• Paper plates<br>• Paper napkins, tissues, and paper towels<br>• Paper take-out boxes and containers | **PLANTS:**<br>• All plant debris, including flowers, leaves, weeds, and<br>• Branches<br>• Tree trimmings (less than 6" in diameter and 4' long)<br><br>**OTHER:**<br>• Bags labeled "Compostable" or BPI<br>• Cooking grease: Small amounts can be soaked up with a paper towel and composted.<br>• Corks (no plastic)<br>• Cotton balls, cotton swabs with paper stems<br>• Hair, fur, and feathers (non-synthetic/colored)<br>• Products clearly labeled "Compostable"<br>• Vegetable wood crates (metal wire OK)<br>• Waxed cardboard and paper<br>• Wood: small pieces of clean wood/sawdust (no plywood/<br>• Pressboard/painted/stained/treated)<br>• Wooden chopsticks, coffee stirrers, toothpicks | • Aluminum foil or trays<br>• "Biodegradable" plastic (not labeled "compostable")<br>• Cat litter or animal feces<br>• Ceramic dishware or glassware<br>• Clothing, linens and rags<br>• Cooking oil<br>• Corks – plastic<br>• Diapers<br>• Dirt, rocks or stone<br>• Flower pots or trays<br>• Foil-backed or plastic-backed paper<br>• Glass, metal or plastic not labeled "Compostable"<br>• Soup cartons and juice boxes<br>• Paper milk, juice and other beverage cartons<br>• Liquids or ice<br>• Plastic bags, wrappers or film<br>• Recyclable/clean cardboard or paper<br>• Styrofoam<br>• Wood – plywood, press board, painted or stained wood |

*Fig. 51. What you can and cannot compost*

## COMPONENT 2: PROCESSING

Processing is the next stage of the process, and it can vary quite a bit depending on the scale of the composting operation, but here's how it works at South Valley Organics.

After they take out all that stuff that can't be composted from the organic waste they receive, svo grinds up the waste using an industrial-sized grinder, making it the ideal size for microorganisms to do their job and decompose the organic waste.

*Fig. 52. The industrial-sized grinder*

Then, the resulting product is placed into long piles called windrows.

*Fig. 53. Windrows*

One thing that really surprised me when I visited South Valley Organics was how hot the compost gets; as a matter of fact, you could even sometimes see steam coming from the piles.

*Fig. 54. Steam coming off of piles*

That's because their operating temperature can get between 140°F (60°C) and 149°F (65°C). Why is that?

Well, 80% to 90% of all microorganisms found in those compost piles are bacteria, and those bacteria can be divided into two groups: aerobic and anaerobic.

Anaerobic bacteria — or bacteria which do not require oxygen — are *sort of* useless and give compost any of the bad smells you might usually associate with compost, but aerobic bacteria, or bacteria which require oxygen levels of at least five percent, are the most efficient and important bacteria in a compost pile. As a matter of fact, if a compost pile doesn't have oxygen (and, thus, doesn't have aerobic bacteria), decomposition can slow by as much as 90%.

Aerobic bacteria can consume basically anything, but love turning nitrogen into protein (which they can use to grow and reproduce) and carbon into energy (which they do by oxidizing the carbon in the organic waste). This process of oxidation generates a lot of heat, but it's a little more nuanced than that.

See, there are different types of aerobic bacteria that show up at different points in the composting process. Psychrophilic bacteria kick into action between 55°F (12.8°C) and 70°F (21.1°C) and produce enough heat so that mesophilic bacteria (which function between 70°F (21.1°C) and 100°F (31.8°C)) can thrive. Note that by the time mesophilic bacteria kick in, the temperature is so hot that the psychrophilic bacteria die off. Mesophilic bacteria are super efficient decomposers and also produce a fair amount of heat, so the whole process repeats when mesophilic bacteria heat up the pile enough for thermophilic bacteria (which function between 113°F (45°C) and 160°F (71.1°C)) to take over and mesophilic bacteria to die off.

You can think of this whole process as a relay race. One bacteria heats up the pile enough for another bacteria to take over and that first bacteria to die, and then that happens again until the pile reaches its ideal temperature of between 140°F (60°C) and 149°F (65°C).

If the temperature gets hotter than 160°F (71.1°C), it can negatively affect the amount of aerobic bacteria and, thus, the productivity of the decomposition, so compost piles must be turned and exposed to air in order to cool them down. Bonus points: doing that exposes the pile to oxygen, which, as we all know, aerobic bacteria needs to operate. And temperature isn't the only condition that needs to be ideal; we also have to make sure the compost pile has enough water so that it's moist but not sopping wet and the ratio of carbon to nitrogen in the pile is somewhere between 25:1 and 30:1.

*Fig. 55.* Turning the compost piles

By the way, if 80% to 90% of all microorganisms found in those compost piles are bacteria, what makes up the other 10% to 20%? Well, fungi like yeasts and molds do. My personal favourite fungi are actinomycetes, which are technically bacteria but operate like fungi and are responsible for that nice earthy smell compost has.

Check that out: in learning about why compost gets so hot, we also learned about the bacteria which power compost and how compost works. Huh, it's almost like I planned that!

On top of those microorganisms, there are a number of macroorganisms that aid in the compost process, including everything from ants and flies to worms and those little rolly polly bugs.

After some time between a couple weeks and a couple months, the compost is finally done. It's refined (using the big ol' machine in figure 56) creating super pure compost and then sent to a lab for testing.

*Fig. 56. Refining the compost*

Testing for what, you might ask? While pathogens like salmonella and weed seeds die at the incredibly high temperatures compost piles operate at, it's always possible for those things to slip through the cracks, so laboratories confirm that each compost pile is free of pathogens. Once they get the okay, South Valley Organics can begin selling the compost back to farmers in the area and the job is done.

So, those are some pretty robust logistics, but do they actually work everywhere?

Well, New York City rolled out it's composting program in 2017. It's currently voluntary, and the advertising around the program is minimal, so they don't have anywhere close to 100% participation. There's certainly the potential for the program to be profitable between the environmental benefits and how it can decrease the cost of exporting trash to landfills, but the program made only $58,000 in 2017 selling compost to farmers, compared to the well over $15 million cost of the program itself. There are currently no plans to expand the program.

Well, that certainly isn't inspiring. However, don't give up hope just yet! San Francisco sends less trash to the landfill than any other major U.S. city, and it's in part because the compost program in San Francisco is thriving. So we know compost programs can work somewhere, but, what's so special about San Francisco?

# Part 4:
# What makes San Francisco's compost program so great?

*Or, the history of Recology and trash collection in San Francisco.*

Not many seventeen-year-olds spend their time reading obscure city ordinances on the subject of sustainability, but that's a pretty typical Tuesday for me.

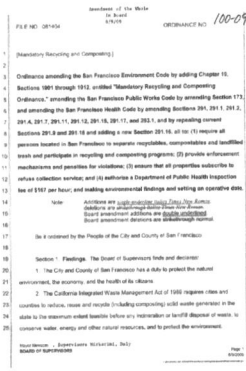

*Fig. 57. The description for this image seems somewhat futile because it will be explained in the following sentence.*

Figure 57 is the first page of a 27-page document called the San Francisco Mandatory Recycling and Composting Ordinance. It was passed back in 2009, and it requires all San Franciscans to recycle and compost by sorting their garbage into recyclables, organic waste, and classic trash. This was completely unprecedented at the time because it was America's first mandatory composting law. Is this what makes San Francisco's program so great? Yes and no. Yes, because this is an incredibly important moment in the history of composting in San Francisco, but no because it doesn't take into account the history of trash collection in the city. And that history starts in the late 1800s.

In the late 19th and early 20th-century, trash collection in San Francisco was mainly done by a group of Italian immigrants who would go around in horse-drawn carriages and collect trash like wood, metal, glass, and organic waste to resell to other people. However, in 1921, the city of San Francisco began actually caring about how their trash was dealt with and scavengers became organized into two associations: the Scavenger's Protective Association and the Sunset Scavenger Company. In 1932, they were both granted licenses to collect SF trash and, in 1935, they decided to merge into a singular company that is, today, known as Recology, the company behind waste management in San Francisco and most of the surrounding Bay Area.

*Fig. 58. Me behind a Recology organic waste bin*

This is one reason San Francisco's compost program is so amazing: their exclusive partnership with Recology. While cities like New York have hundreds of companies competing to collect waste, San Francisco only works with Recology, decreasing administrative and logistical friction, which, in turn, means it's a lot easier to try new things.

Indeed, in the past, San Francisco and Recology have run many "pilot programs" (which are small scale tests of larger projects done to see if those projects would be successful). Many of these pilot programs turned out really well. In 1996, the curbside collection of food scraps from San Francisco residents was pilot tested, and it went so well that the project was rolled out to the whole city in 2001. Moreover, in 1999, "the Fantastic Three" program was pilot tested. The Fantastic Three are the three garbage bins every resident and business located in San Francisco has: blue for recyclables, black for landfill-bound trash, and green for compostables.

*Fig. 59.* The Fantastic Three

While Recology really does value sustainable waste management — their motto, "Waste Zero," can be found on all of their garbage cans — the genius of this whole system is that it's structured so that Recology has strong financial incentives to divert waste from landfills. Recology has a lot of stake in various recycling and composting facilities around the Bay Area, but they don't have stake in landfills, so it's to their benefit to move as much waste to recycling and composting plants as possible.

On top of Recology, another reason San Francisco's compost program is so great is that, quite simply, the government wants it; composting has the full support of the San Francisco government.

It all started in 1989 with the state government. The California Assembly passed the Integrated Waste Management Act, which set the goal of a diversion rate of 25% by 1995 and 50% by 2000. A diversion rate is the measure of how much trash you're diverting from landfills by recycling or composting that trash instead. At the time, San Francisco's diversion rate was 10%, which means it was sending 90% of its trash to landfills.

Over the next decade, San Francisco invested a ton of money into making their waste management system as environmentally friendly as possible, and shortly after they introduced the Fantastic Three program, they hit their target of a 50% diversion rate. But San Francisco aspired to do even better. In 2002, SF's Board of Supervisors passed the Zero Waste Goal, calling for San Francisco to have a diversion rate of 75% by 2010 and 100% by 2020.

That's right: San Francisco wants to send no trash to landfills. And they're actually kind of doing it.

As a result of extensive legislation, in 2007, the city reached a diversion rate of 72%, while the state of California was hovering at 52%. But San Francisco realized that without a mandatory curbside food waste collection program that involved people sorting their waste into recyclables, compost, and trash, they wouldn't be able to stay on target and reach their goal of zero waste. And that's where the 2009 Mandatory Recycling and Composting Ordinance comes in; thanks to that decree, between its passing in 2009 and 2011, composting in SF increased by 45%, and in 2018, the city hit a diversion rate of 80%. That means over 1.5 million tons of garbage are diverted from landfills every year.

So, no, San Francisco doesn't send no waste to landfills, but they've gotten pretty darn close (at least, closer than any other major US city), and that certainly means they're doing something right.

Today, the city continues to work hard at getting their diversion rate higher. The San Francisco Department of the Environment has an entire team dedicated to raising awareness about San Francisco's zero waste goals, which includes doing everything from providing assistance when people request it to going door to door and informing residents about any rule changes related to waste management. Moreover, they give $600,000 a year to local nonprofits making innovations in the waste management sector.

So, what's the future for compost in San Francisco? Well, if everyone were to fully abide by the Mandatory Recycling and Composting Ordinance and separate their trash, SF would be able to achieve a diversion rate of 90%, so the city is continuing to work towards that and find ways to get that last 10%. But the future looks bright, because the residents are on its side. Before this law was enacted, the city surveyed a bunch of apartment owners and found that 85% of them really liked the program, and when was the last time that many people agreed on anything political?

Obviously, there is a lot we need to do to fight climate change, and even that's an understatement. But one factor we need to look at is what we do with our food and what we do with our waste. And compost is the brilliant solution to both of those things.

Now that we've changed the disconnectedness we have with the stuff we throw away, you have the ability to change that disconnectedness in others as well. You can encourage them to care about a brilliant solution which would lead to a greener, better future for ourselves and our kids and our grandkids.

In the words of Walt Whitman, "Behold this compost! behold it well!"

*Fig. 60. A double rainbow I found on the Mountain School's campus*

# Part 5: An Appendix

*Sources, Image Citations, Special Thanks, and About the Author*

### SOURCES

Sources are organized by topic and ranked (roughly) in order of importance. The sources include many links, so if you would like an online version of this list, it is available at beholdthiscompost.com/sources.

**The Donner Party**

- KnowingBetter. (2017, November 22). You Have Died of... Cannibalism | Oregon Trail. Retrieved from https://www.youtube.com/watch?v=k-aA3FtE1QE.

**Why Waves Break White**

- Boon-Ying, Lee. (n.d.). Why do breaking waves appear white? Hong Kong Observatory. Retrieved from https://www.hko.gov.hk/m/article_e.htm?title=ele_00361.

**The Mountain School and Its History**

- (n.d.). The Mountain School Website. Retrieved from https://www.mountainschool.org.

- (2013, May 17). Waller MacNiven Conard. Valley News. Retrieved from https://www.vnews.com/Archives/2013/05/Waller-Conard-obit-vn-051713.

**The Newtown Pippin**

- Jacobsen, R. (2014). Apples of Uncommon Character: Heirlooms, Modern Classics, and Little-Known Wonders. New York, NY: Bloomsbury.

- Karp, D. (2003, November 5). It's Crunch Time for the Venerable Pippin. The New York Times. Retrieved from https://www.nytimes.com/2003/11/05/dining/it-s-crunch-time-for-the-venerable-pippin.html.

- Hitchcock, S. T. (1982, October 10). At the Core Of Apple Appeal. The Washington Post. Retrieved from https://www.washingtonpost.com/archive/lifestyle/food/1982/10/10/at-the-core-of-apple-appeal/f112c6ab-9efb-4848-bc96-9767c1668f0e/.

- Beach, S. A. (1905). The Apples Of New York.

- The Editors of Britannica. (2019, January 4). Graft. Encyclopedia Britannica. Retrieved from https://www.britannica.com/topic/graft.

- (2016, December 8). New Census Data Show Differences Between Urban and Rural Populations. The United States Census Bureau. Retrieved from https://www.census.gov/newsroom/press-releases/2016/cb16-210.html.

**Food Waste**

- Reich, A. H. (2015, August). Demand-Side Approaches to Improving Global Food Sustainability (dissertation). The University of Minnesota. Retrieved from https://conservancy.umn.edu/bitstream/handle/11299/174833/Reich_umn_0130M_16231.pdf;sequen.

- Lipinski, B., Hanson, C., Lomax, J., Kitinoja, L., Waite, R., & Searchinger, T. (2013). Reducing Food Loss and Waste. World Resources Institute. Retrieved from http://pdf.wri.org/reducing_food_loss_and_waste.pdf.

- Gustavsson, J., Cederberg, C., Sonesson, U., Otterdijk, R. van, & Meybeck, A. (2011). Global Food Losses and Food Waste. The Food and Agriculture Organization of the United Nations. Retrieved from http://www.fao.org/3/mb060e/mb060e00.htm.

- Reich, A. H. (2014, April). Food Loss and Waste in the US: The Science Behind the Supply Chain. The University of Minnesota. Retrieved from https://www.cahfs.umn.edu/sites/cahfs.umn.edu/files/brief_food-loss-waste_2018.pdf.

- Hall, K. D., Guo, J., Dore, M., & Chow, C. C. (2009, November 25). The progressive increase of food waste in America and its environmental impact. PLoS One. Retrieved from https://www.ncbi.nlm.nih.gov/pubmed/19946359.

- Gunders, D. (2009, August). Wasted: How America Is Losing Up to 40 Percent of Its Food from Farm to Fork to Landfill. Natural Resources Defense Council. Retrieved from https://www.nrdc.org/sites/default/files/wasted-food-IP.pdf.

- (n.d.). Sustainable Management of Food Basics. The Environmental Protection Agency.

Retrieved from https://www.epa.gov/sustainable-management-food/sustainable-management-food-basics.

- (n.d.). Inventory of U.S. Greenhouse Gas Emissions and Sinks: 1990-2011. The Environmental Protection Agency. Retrieved from https://www.epa.gov/sustainable-management-food/sustainable-management-food-basics.

- (n.d.). Inventory of U.S. Greenhouse Gas Emissions and Sinks: 1990-2011. The Environmental Protection Agency. Retrieved from https://www.epa.gov/ghgemissions/inventory-us-greenhouse-gas-emissions-and-sinks-1990-2011.

- Elkins, K. (2018, January 29). US families waste $1,500 a year throwing out food—here's how to save more and eat better. CNBC. Retrieved from https://www.cnbc.com/2018/01/29/families-waste-1500-a-year-on-food-save-by-making-meals-from-scraps.html.

**Global Climate**

- Ruddiman, W. F. (2000, December 15). Earth's Climate: Past and Future. W. H. Freeman and Company.

- Nutman, A. P., Bennett. V. C., Friend C. R. L., Van Kranendonk, M. J., Chivas, A. R. (2016, September 22). Rapid emergence of life shown by discovery of 3,700-million-year-old microbial structures. Nature. Retrieved from https://www.nature.com/articles/nature19355.

**Plate Tectonics**

- Raymo, K. (1983, November 1). The Crust of Our Earth: An Armchair Traveler's Guide to the New Geology. Prentice Hall Direct.

- Raymo, K., Raymo, M. E. (2001, February 10). Written in Stone: A Geological History of the Northeastern United States. Black Dome Press.

- McCoy, R. M. (2006, June 22). Ending in Ice: The Revolutionary Idea and Tragic Expedition of Alfred Wegener. Oxford University Press.

- Oreskes, N. (2003, February 4). Plate Tectonics: An Insider's History Of The Modern Theory Of The Earth. Westview Press.

- Ruddiman, W. F. (2000, December 15). Earth's Climate: Past and Future. W. H. Freeman and Company.

- Wegener, A. (1966). The Origin of Continents and Oceans. (J. Biram, Trans.). Dover Publications.
- DeConto, R. M. (2008, May 23). Plate Tectonics and Climate Change. The University of Massachusetts at Amherst. Retrieved from https://www.geo.umass.edu/climate/papers2/deconto_tectonics&climate.pdf.

**Milankovitch Cycles**

- Ruddiman, W. F. (2010, April 11). Plows, Plagues, and Petroleum: How Humans Took Control of Climate. Princeton University Press.
- (n.d.). Milankovitch Cycles. The State Climate Office of North Carolina. Retrieved from https://climate.ncsu.edu/edu/Milankovitch.

**Anthropogenic Climate Change**

- (2016, August). Climate Change: Science and Impacts Factsheet. The University of Michigan. Retrieved from http://css.umich.edu/factsheets/climate-change-science-and-impacts-factsheet.
- (2016, August). Greenhouse Gasses Factsheet. The University of Michigan. Retrieved from http://css.umich.edu/factsheets/greenhouse-gases-factsheet.
- (n.d.). World of Change: Global Temperatures. NASA's Earth Observatory. Retrieved from https://earthobservatory.nasa.gov/world-of-change/DecadalTemp.
- (2001). Climate Change Science: An Analysis of Some Key Questions. The National Research Council. Retrieved from https://www.nap.edu/catalog/10139/climate-change-science-an-analysis-of-some-key-questions.
- (2006). Surface Temperature Reconstructions for the Last 2,000 Years. The National Research Council. Retrieved from https://www.nap.edu/catalog/11676/surface-temperature-reconstructions-for-the-last-2000-years.

**How City-Wide Compost Programs Work**

- (n.d.). The Composting Process. The University of Illinois. Retrieved from https://web.extension.illinois.edu/compost/process.cfm.
- Ross, R. (2018, September 12). The Science Behind Composting. LiveScience. Retrieved from https://www.livescience.com/63559-composting.html.

- Trautmann, N. (1996). Compost Physics. Cornell University's Waste Management Institute. Retrieved from http://compost.css.cornell.edu/physics.html.

- Bement, L. (2009, December). Hot Composting vs. Cold Composting. Fine Gardening. Retrieved from https://www.finegardening.com/article/hot-composting-vs-cold-composting.

- Rueb, E. S. (2017, June 2). How New York Is Turning Food Waste Into Compost and Gas. The New York Times. Retrieved from https://www.nytimes.com/2017/06/02/nyregion/compost-organic-recycling-new-york-city.html.

- Collins, L. M. (2018, November 9). The Pros and Cons of New York's Fledgling Compost Program. The New York Times. Retrieved from https://www.nytimes.com/2018/11/09/nyregion/nyc-compost-zero-waste-program.html.

- (n.d.). What Goes Where. Recology. Retrieved from https://www.recology.com/recology-san-francisco/what-goes-where/.

### What makes San Francisco's compost program so great?

- Tam, L. (2010, February 1). Toward Zero Waste: A look at San Francisco's model recycling policies. The San Francisco Bay Area Planning and Urban Research Association. Retrieved from https://www.spur.org/publications/urbanist-article/2010-02-01/toward-zero-waste.

- Howard, B. C. (2013, June 18). How Cities Compost Mountains of Food Waste. National Geographic. Retrieved from https://www.nationalgeographic.com/news/2013/6/130618-food-waste-composting-nyc-san-francisco/.

- Brigham, K. (2018, July 14). How San Francisco sends less trash to the landfill than any other major U.S. city. CNBC. Retrieved from https://www.cnbc.com/2018/07/13/how-san-francisco-became-a-global-leader-in-waste-management.html.

- Cherney, M. (2014, March 24). How to Turn Compost into Cash. Vice News. Retrieved from https://www.vice.com/en_us/article/539gwb/how-to-turn-compost-into-cash.

- (n.d.). Mayor Lee Announces San Francisco Reaches 80 Percent Landfill Waste Diversion, Leads All Cities in North America. The San Francisco Department of the Environment. Retrieved from https://sfenvironment.org/news/press-release/mayor-lee-announces-san-francisco-reaches-80-percent-landfill-waste-diversion-leads-all-cities-in-north-america.

- Cheng, E. (2014, December 23). Are you gonna eat that? The future of recycling. CNBC. Retrieved from https://www.cnbc.com/2014/12/22/composting-may-be-future-of-recycling-with-us-cities-leading-the-way.html.

- McClellan, J. (2017, August 3). How San Francisco's mandatory composting laws turn food waste into profit. The Arizona Republic. Retrieved from https://www.azcentral.com/story/entertainment/dining/food-waste/2017/08/03/san-francisco-mandatory-composting-law-turns-food-waste-money/440879001/.

- Charles, J. (2018). San Francisco closes the lid on garbage. Plenty Magazine. Retrieved from https://www.mnn.com/lifestyle/recycling/stories/san-francisco-closes-the-lid-on-garbage.

- (2009, June 9). San Francisco's Mandatory Recycling and Composting Ordinance. The City and County of San Francisco. Retrieved from https://sfenvironment.org/sites/default/files/policy/sfe_zw_sf_mandatory_recycling_composting_ord_100-09.pdf.

## IMAGE CITATIONS

*Fig. 1.* Hotshot Imaging

*Fig. 2.* Alexander Nickel

*Fig. 3.* Alexander Nickel

*Fig. 4.* Alexander Nickel

*Fig. 5.* Alexander Nickel

*Fig. 6.* Alexander Nickel

*Fig. 7.* Alexander Nickel

*Fig. 8.* Alexander Nickel

*Fig. 9.* Alexander Nickel

*Fig. 10.* Alexander Nickel

*Fig. 11.* Alexander Nickel

*Fig. 12.* "Malus domestica: Newtown Pippin" by Ellen Isham Schutt is licensed under public domain; https://commons.wikimedia.org/wiki/File:Pomological_Watercolor_POM00000792.jpg

*Fig. 13.* "Newtown pippins" by Leslie Seaton is licensed under CC BY 2.0; https://commons.wikimedia.org/wiki/File:Newtown_pippins_(8167963860).jpg

*Fig. 14.* "Apple juice glass bottles in market" by Marco Verch is licensed under CC BY 2.0; https://www.flickr.com/photos/30478819@N08/42708791441

*Fig. 15.* "Earthrise" by NASA/Bill Anders is licensed under public domain; https://commons.wikimedia.org/wiki/File:NASA-Apollo8-Dec24-Earthrise.jpg#/media/File:NASA-Apollo8-Dec24-Earthrise.jpg

*Fig. 16.* "Horodyskia fossils, about 1.4 billion years old – Redpath Museum – McGill University – Montreal, Canada" by Daderot is licensed under CC0 1.0; https://commons.wikimedia.org/wiki/File:Horodyskia_fossils,_about_1.4_billion_years_old_-_Redpath_Museum_-_McGill_University_-_Montreal,_Canada_-_DSC07903.jpg

*Fig. 17.* "Map Of Major Tectonic Plates In The World" by Blatant World is licensed under CC BY 2.0; https://www.flickr.com/photos/blatantworld/5051807235

*Fig. 18.* Alexander Nickel

*Fig. 19.* "Oceanic spreading" by Surachit is licensed under CC BY-SA 3.0; https://commons.wikimedia.org/wiki/File:Oceanic_spreading.svg

*Fig. 20.* "Continental-continental constructive plate boundary" by Domdomegg is licensed under CC BY 4.0; https://commons.wikimedia.org/wiki/File:Continental-continental_constructive_plate_boundary.svg

*Fig. 21.* "Continental-continental conservative plate boundary opposite directions" by Domdomegg is licensed under

CC BY 4.0; https://commons.wikimedia.org/wiki/File:Continental-continental_conservative_plate_boundary_opposite_directions.svg

*Fig. 22.* "Oceanic-continental destructive plate boundary" by Domdomegg is licensed under CC BY 4.0; https://commons.wikimedia.org/wiki/File:-Oceanic-continental_destructive_plate_boundary.svg

*Fig. 23.* "Alfred Wegener ca.1924-30" by Bildarchiv Foto Marburg is licensed under public domain {{PD-US-expired}}; https://commons.wikimedia.org/wiki/File:Alfred_Wegener_ca.1924-30.jpg

*Fig. 24.* "Pangea political" by Massimo Pietrobon is licensed under CC BY 3.0; https://commons.wikimedia.org/wiki/File:Pangea_political.jpg

*Fig. 25.* "seafloor spreading" by Ramona Benson is licensed under CC BY 2.0; https://www.flickr.com/photos/114042825@N07/11877403915

*Fig. 26.* "Oceanic.Stripe.Magnetic.Anomalies.Scheme" by Chmee2 is licensed under public domain; https://commons.wikimedia.org/wiki/File:Oceanic.Stripe.Magnetic.Anomalies.Scheme.svg

*Fig. 27.* "Milankovitch-cycles hg" by Hannes Grobe is licensed under CC BY-SA 2.5; https://commons.wikimedia.org/wiki/File:Milankovitch-cycles_hg.png

*Fig. 28.* "Perihelion-Aphelion" by Chris55 is licensed under CC BY-SA 4.0; https://commons.wikimedia.org/wiki/File:Perihelion-Aphelion.svg

*Fig. 29.* "Urbain Le Verrier" by Magnus Manske is licensed under public domain {{PD-US-expired}}; https://commons.wikimedia.org/wiki/File:Urbain_Le_Verrier.jpg

*Fig. 30.* "Earth precession" by NASA, Mysid is licensed under public domain; https://commons.wikimedia.org/wiki/File:Earth_precession.svg

*Fig. 31.* "Jean d'Alembert" by Quentin de La Tour is licensed under public domain; https://fr.wikipedia.org/wiki/Fichier:-Jean_d%27Alembert.jpeg

*Fig. 32.* "Orbital relations of the Solstice, Equinox & Intervening Seasons" by Colivine is licensed under CC0 1.0; https://commons.wikimedia.org/wiki/File:Orbital_relations_of_the_Solstice,_Equinox_%26_Intervening_Seasons.svg

*Fig. 33.* "Earth obliquity range" by NASA, Mysid is licensed under public domain; https://commons.wikimedia.org/wiki/File:Earth_obliquity_range.svg

*Fig. 34.* "The Sun by the Atmospheric Imaging Assembly of NASA's Solar Dynamics Observatory – 20100819" by NASA/SDO (AIA) is licensed under public domain; https://en.wikipedia.org/wiki/File:The_Sun_by_the_Atmospheric_Imaging_Assembly_of_NASA%27s_Solar_Dynamics_Observatory_-_20100819.jpg

*Fig. 35.* "Solar irradiance and temperature 1880-2018" by NASA is licensed under public domain; https://commons.wikimedia.org/wiki/File:Solar_irradiance_and_temperature_1880-2018.jpeg

*Fig. 36.* "Atmospheric carbon dioxide concentrations and global annual average temperatures over the years 1880 to 2009" by the United States Global Change Research Program, NOAA/NCDC, Thomas R. Karl, Jerry M. Melillo, and Thomas C. Peterson is licensed under public domain; https://commons.wikimedia.org/wiki/File:Atmospheric_carbon_dioxide_concentrations_and_global_annual_average_temperatures_over_the_years_1880_to_2009.png

*Fig. 37.* "Global Carbon Cycle" by the Atmospheric Infrared Sounder is licensed under CC BY 2.0; https://www.flickr.com/photos/atmospheric-infrared-sounder/8263952221

*Fig. 38.* "Global Carbon Emissions" by Mak Thorpe and Autopilot is licensed under CC BY-SA 3.0; https://commons.wikimedia.org/wiki/File:Global_Carbon_Emissions.svg

*Fig. 39.* "Annual world greenhouse gas emissions, in 2005, by sector" by Enescot is licensed under CC0 1.0; https://commons.wikimedia.org/wiki/File:Annual_world_greenhouse_gas_emissions,_in_2005,_by_sector.svg

*Fig. 40.* Alexander Nickel

*Fig. 41.* "Maps of the 2013 global temperature anomaly (top) and the 1950-2013 temperature trend (bottom.)" by NASA/GSFC/Earth Observatory, NASA/GISS is licensed under public domain; https://www.giss.nasa.gov/research/news/20140121/

*Fig. 42.* "Global warming – change in total heat content of earth" by Skeptical Science is licensed under CC BY 3.0; https://commons.wikimedia.org/wiki/File:Global_warming_-_change_in_total_heat_content_of_earth.jpg

*Fig. 43.* "The Rising Sea Level" by NASA is licensed under public domain; https://commons.wikimedia.org/wiki/File:The_Rising_Sea_Level.jpg

*Fig. 44.* "Eastont2" by Pelto is licensed under public domain; https://commons.wikimedia.org/wiki/File:Eastont2.jpg

*Fig. 45.* "Cute puppy" by kitty.green66 is licensed under CC BY-SA 2.0; https://www.flickr.com/photos/53887959@N07/4985420598

*Fig. 46.* https://pixabay.com/photos/fresh-compost-hand-man-2386786/

*Fig. 47.* Alexander Nickel

*Fig. 48.* Alexander Nickel

*Fig. 49.* Alexander Nickel

*Fig. 50.* "Cow female black white" by Keith Weller/USDA is licensed under public domain; https://commons.wikimedia.org/wiki/File:Cow_female_black_white.jpg

*Fig. 51.* Alexander Nickel

*Fig. 52.* Alexander Nickel

*Fig. 53.* Alexander Nickel

*Fig. 54.* Alexander Nickel

*Fig. 55.* Alexander Nickel

*Fig. 56.* Alexander Nickel

*Fig. 57.* https://sfenvironment.org/sites/default/files/policy/sfe_zw_sf_mandatory_recycling_composting_ord_100-09.pdf

*Fig. 58.* Alexander Nickel

*Fig. 59.* Alexander Nickel

*Fig. 60.* Alexander Nickel

## SPECIAL THANKS

Steven and Tammy Nickel

South Valley Organics

Izzy Swart

Tyler Poon

Clay Adams

Billy Pierce

Evan Serre

Andres Ibarra

Elise Watt

Sam Kelman

Ben Tiefenthaler

Kathy Hooke

Emily Sartin

Luke De

Dave Wiskus

Brick Treybig

Simon Buckmaster

Gemma Arnott

Everyone at Standard, the Mountain School, and the Nueva School who supported me along the way. While the people listed above helped with the book directly, many people helped me indirectly. Thank you all so incredibly much.

## ABOUT THE AUTHOR

Alexander Ulysses Nickel is the creator of Technicality (www.youtube.com/technicality), an educational YouTube channel about science, humanities, and anything he finds fascinating. Alex started Technicality around 6 years ago when he was studying blended learning and using educational videos in the classroom; his goal was to create content that communicates science in an engaging way. Since then, he has researched, scripted, filmed, hosted, edited, and published over 70 episodes on everything from psychology to history to physics. Technicality has gained over 50,000 subscribers and millions of total views. He has been written about on websites like Huffington Post and Mashable, given a TEDx Talk, and worked with YouTubers such as Physics Girl, Tom Scott, Counter Arguments, and many others. Technicality videos have been shown in middle school, high school, and university classrooms all over the world. This is his first book.

When he's not working on Technicality, Alex is a high school student, the co-chair of his city's youth commission, a content and curriculum intern for Education.com, the lead role in his school's production of the play Harvey, the co-creator of the educational subscription streaming service Nebula (watchnebula.com), and a prolific mountain biker and outdoorsman. Alex lives in the San Francisco Bay Area with his mother Tammy, his father Steven, and his dog Blue.

www.ingramcontent.com/pod-product-compliance
Lightning Source LLC
Chambersburg PA
CBHW041352290426
44108CB00001B/18